SPOOKY PHYSICS

MSAC Philosophy Group | 2016

This is an illustrated version of Dr. Andrea Diem-Lane's book, *Spooky Physics*, which has also been republished under the title *Quantum Weirdness*.

Einstein vs. Bohr

Introduction: An Age-Old Problem

What is truly real? And how do we know? These twin questions, sometimes related in philosophical jargon to ontology and epistemology, are of central importance in understanding the dramatic implications of quantum theory. Indeed, one could argue that the reason quantum theory is so baffling to understand is because it upends our deepest and most cherished ideas about what is real and what is not. Moreover, quantum theory calls into question the very process of how we know things. It is for this reason that Albert Einstein resisted the implications of quantum theory because he knew that what it portended was an end to determinism and an end to a strict causality governing the universe. Of course, for others like Niels Bohr, succumbing to such indeterminism, even if intellectually disagreeable, is precisely what the theory demands. In other words, it is indeterminism itself which informs all of quantum theory, and to neglect that marked characteristic is to ignore its most vital feature.

Einstein ultimately found the implications of quantum theory so unsettling that he made a number of terse remarks on it. In a letter to Max Born, dated September 1944, he wrote, "You believe in the God who plays dice, and I in complete law and order in a world which objectively exists, and which I, in a wildly speculative way, am trying to capture. I hope that someone will discover a more realistic way, or rather a more tangible basis than it has been my lot to find. Even the great initial success of the Quantum Theory does not make me believe in the fundamental dice-game…." In fact, Einstein spent a good sum of his life trying to come up with thought experiments which would demonstrate the incompleteness of quantum theory and show why it was at best an interregnum theory which would in time yield to a more reasonable and deterministic one. As he quipped to Born, "Although I am well aware that our younger colleagues interpret this as a consequence of senility. No doubt the day will come when we will see whose instinctive attitude was the correct one." This book is a brief introduction to the famous Einstein-Bohr debate over the implications of quantum theory with a special focus on the philosophical ramifications of Heisenberg's uncertainty principle.

We are fortunate that there exists a fairly extensive record of the conversations between these two eminent thinkers. Indeed, it isn't hyperbolic to call the Einstein-Bohr conflict one of the greatest intellectual debates of modern times, nay of any time period. This book begins with an overview of quantum theory and its early development. It also explores some of its weirder aspects, including the dual aspect of light quanta. In Chapter two, we explore why Einstein found many aspects of quantum theory so disagreeable, especially the idea of uncertainty relations where knowing an electron's position increases the ignorance of knowing an electron's momentum, and vice versa.

3 | Spooky Physics

Chapter three centers on why Bohr accepted quantum indeterminism (what he called complimentarity) and encouraged his colleagues and students (such as Pauli and Born and Heisenberg) to play out its consequences to the fullest—what would later be famously called the Copenhagen school. The most heated section of the book comes in Chapter four where we get to witness (both through transcriptions of talks given at the time and through extensive correspondence, particularly the letters to and from Max Born) the passion of Einstein's arguments against quantum theory and Bohr's equal passion for it. Although both Bohr and Einstein have been dead for decades, Chapter five illustrates that their debate still lives on and why it is still a very hot topic even among a newer generation of physicists. And, finally, in the conclusion we ask what this debate means for us and our day to day lives.

Our evolution has bounded what we can and cannot know about the world around us. Because of this our brains are not well adapted to understand either the very large or the very small. We are quite literally middling creatures that have been shaped for eons of time to survive in eco-niches where our food and prey are accessible to our five apertures. What this means, of course, is that whenever we venture beyond our middle earth by extending our senses to the very large or very small, we have to acclimate ourselves anew.

The history of science is a record of how humans achieved such acclimations and how, in turn, such new insights transformed our understanding of how the universe actually works. Whether it was Galileo's telescope (seeing a pock marked moon versus a polished lunar surface) or Copernicus mathematical equations (indicating a solar based orbital system versus an earth centered one), in each case sensory or mental breakthroughs led to revolutions in scientific thought. It may be no exaggeration to say that whenever we altered our bodily or cranial limits we extended our world, a world which is forever linked to the limitations of what the senses can and cannot reveal.

To say that neurology is ontology is merely to state the obvious. But what sometimes gets lost in such clichés is that our brain state is never static and thus the world is never the same as well. Change the neural apparatus and one transforms the universe. Not necessarily because the brain creates such realities, but rather because the limitations of one's cranial capacities predetermines what is accessible or knowable about any given aspect of reality. Change those neural coordinates and thereby change one's intellectual map.

Spooky Physics

All of this is necessary preface to understand why the human mind has an almost innate difficulty in understanding quantum theory—a theory which takes into account things so infinitesimally tiny that even our best analogies freeze our minds in a state of wonder.

Ludwig Wittgenstein gives us a fruitful glimpse of just how contradictory quantum physics can be and why it demonstrates prima facie its almost inherent illogical nature. Writing several years before the discovery of Werner Heisenberg's Nobel Prize winning discovery of the uncertainty relations in the subatomic realm, Wittgenstein states in his famous *Tractatus Logico-Philosophicus*:

"6.3751: For two colours, e.g. to be at one place in the visual field, is impossible, logical impossible, for it is excluded by the logical structure of colour. Let us consider how this contradiction presents itself in physics. Somewhat as follows: That a particle cannot at the same time have two velocities, i.e. that at the same time it cannot be in two places, i.e. that particles in different places at the same time cannot be identical. It is clear that the logical product of two elementary propositions can neither be a tautology nor a contradiction. The assertion that a point in the visual field has two different colours at the same time is a contradiction."

Today, of course, quantum physicists state the opposite of Wittgenstein's logical necessity about the behavior of matter and point out that indeed a particle can be in two places at the same time, even if that space and time is limited in its regional and temporal import.

What Wittgenstein captured (quite unwittingly since his Tractatus dates from the latter part of the First World War) was how a rational, logical mind would be upended by the implications of quantum theory. Moreover, he provides us with a framework for why it may be so difficult for many of us to actually "get" quantum theory. As Richard Feynman, the well-known architect behind Quantum Electrodynamics (QED), once quipped, "I think I can safely say that nobody understands quantum mechanics."

I can think of no better caveat than Feynman's when approaching this most profound of subjects. Listening in on the Einstein-Bohr debate may not resolve our own existential dilemmas, but it will undoubtedly put into sharp relief what is at stake when confronting the heart of matter itself.

Quantum Weirdness

Imagine taking a road trip to Las Vegas, Nevada, from Huntington Beach, California. Depending on the traffic, and how fast one is driving, it may take anywhere from five to eight hours. However, there are sections along the way (particularly near Barstow) where speeding too fast will most likely result in being stopped by the highway patrol and receiving a ticket. This can happen even when there appears to be no law enforcement officers in sight.

Why? Because of sophisticated radar tracking stations that monitor traffic flow. Radio waves emitted from the station spread out in varying directions and when they hit a moving object, some of those waves bounce back and are received again at the tracking station, containing two key pieces of information: the position of the vehicle and its momentum. This may seem a bit trivial but these forms of information are absolutely vital to understanding almost anything in the physical universe. Indeed, one can exaggerate here a bit and say almost all of physics is based on these two points of data. Knowing only the position of the car, for instance ("hey, there's a Ford truck in Duarte"), isn't sufficient to warrant a speeding ticket. And if one only knows the momentum of the vehicle but not its whereabouts it is a bit frustrating.

Now this fairly trite example can be applied to almost any event in our day to day lives, from when to attend a lecture at the local college, to when we pick up our children from elementary school, to when we pick up pizza at our neighborhood restaurant. The civilized world is fundamentally based on knowing both the position and momentum of physical objects, including when and where to pick up our spouses from shopping at Target.

Newtonian physics is a picture of this mechanistic and predictable universe and, as such, serves us well in navigating our lives through most events. However, when we begin to look at bits of matter that are extraordinarily small, this same guiding map breaks down.

Imagine now that instead of taking a car to Las Vegas, you are riding on a single electron (to slightly butcher Einstein's more famous metaphor of riding on a beam of light), traveling much faster than the speed limit of 70 miles per hour. Indeed, you would be approaching the ultimate speed that any particle can travel, 186,000 miles per second. Clearly, such speeding warrants a ticket! But, in this instance, the electron police find themselves in a very strange conundrum. Because what they discover to their chagrin is that the more they comprehend how fast the electron is traveling the less they know about where it is exactly located. And, then, when they do get a fix on its position, they lose sight of its momentum.

What they soon realize is that their very act of measuring is interfering with the electron's ultimate position and/or momentum. It is as if the radar itself (which is in truth nothing more than electromagnetic energy) is literally moving the electron at different speeds and/or in different directions.

This is akin to being in a car and having the radar either bump up your speed a hundred miles an hour or having it transport you to another freeway and sending you on your way to San Diego. Something is clearly wrong with this picture. Something is clearly breaking down. And if this happened in our wayward drives to cities in the desert, the highway patrol's radar tracking station would be directly responsible for our speeding violations or for our confused and haphazard sojourns. One could literally say to the ticketing patrolman, "But you made me speed and/or switch lanes. Therefore, you should be giving yourself the ticket, not me." And, given what we know about how the radar interfered with your car, the judge would be forced to admit the obvious and let you off and reprimand the traffic station.

Enter the weird and twisted world of quantum mechanics. While our car example only becomes viable at the level of the very small, it is disconcerting nonetheless to realize that Newtonian physics breaks down precisely when one gets closer to the secrets of Nature. In order to understand why uncertainty increases when we explore the very small, we need to first understand what Max Planck discovered over a century ago when studying black body radiation.Instead of radiant energy being emitted purely in waves and in smaller and smaller frequencies which could be halved ad infinitum, Planck theorized (though apparently he thought that his views would be only a temporary bridge) that energy came in discrete packets, quantified bits of matters, known more popularly later as "quanta." Quanta cannot be halved, and thus electromagnetic energy can only come in multiples of this basic unity of energy, later known famously as Planck's constant.

This is analogous to when one goes to the store to buy a bottle of classic coke. Let's say the 20 ounce bottle costs one dollar and twenty-nine cents (the current price at our local 7/11) and you give the clerk 2 bucks. Now imagine when you get your change of 71 cents that you object to the penny and argue that you want something "smaller" than the penny, like 1/10 of a cent or even a ½ of a cent. The clerk will no doubt look at you a bit strange and he or she may reply, "But we have nothing less than a penny. That's the lowest amount of money available."

You cannot "halve" a penny in our day to day world. Likewise, you cannot halve a quanta. Nature, it seems, has decided that the smallest unit for exchange is this single photon and apparently there is no way around this. You literally cannot "short" change nature, even if your physicist's intuition suggests that you should be able to. Of course Planck's constant is indeed very small, 6.626068 × 10-34 m2 kg / s.

So small, in fact, that our minds cannot really grasp, except with faulty analogies and metaphors, the tininess of the subatomic realm. The implications of the quantification of matter was not lost on Albert Einstein who used Planck's understanding to develop his theory on the photoelectric effect for which he eventually won his only Nobel Prize.

What was so disconcerting about Planck's discovery (or, should we say unintended uncovering) and Einstein's photoelectric effect was that for decades physicists had experimentally demonstrated that light acted like a wave, but now there was evidence of its particle-like nature. This dualistic realization about the nature of matter forced the world of science into a theoretic crises. How can it be both? Or, as the experiments at the time indicated, how is it that in one context light propagates as if it was a wave and in another context light behaves as if it were composed of tiny bullets? Is nature so capricious?

Moreover, if light is both a wave and a particle why is it that only one aspect (but not both) shows up in varying experimental designs? Do we really choose how light is going to behave?

The famous double slit experiment illustrates very clearly the inherent weirdness of the quanta world. Richard Feynman, the famous Nobel Prize winning physicist of the 1960s, has stated that analyzing this experiment alone can reveal the deep mysteries of quantum mechanics.

There have been a large number of books (and even a few films) which have explained how the double slit experiment works. It was first devised by Thomas Young in the early part of the 19th century in which he devised an obstacle with two openings and passed a beam of light through the apertures which would then hit an adjoining barrier wall. What he found was that when light passed through these two slits it caused an interference pattern showing that light had a wave light aspect. However, later experiments showed that if you only had one slit open, light acted as a discrete packet (a quanta) which demonstrated that it had a particle or bullet like aspect.

How the light behaves depends on how the experiment is set-up. Open up just one slit and light acts like a particle. Open up two slits and light acts like a wave. But, the real question (and the one at the heart of quantum weirdness) is how does the light know if the other slit is open or closed? Even if only one photon is allowed to pass through only one slit, if the other slit is open it will act like a wave. If, however, the other slit is closed, the light will be particle like.

8 | Spooky Physics

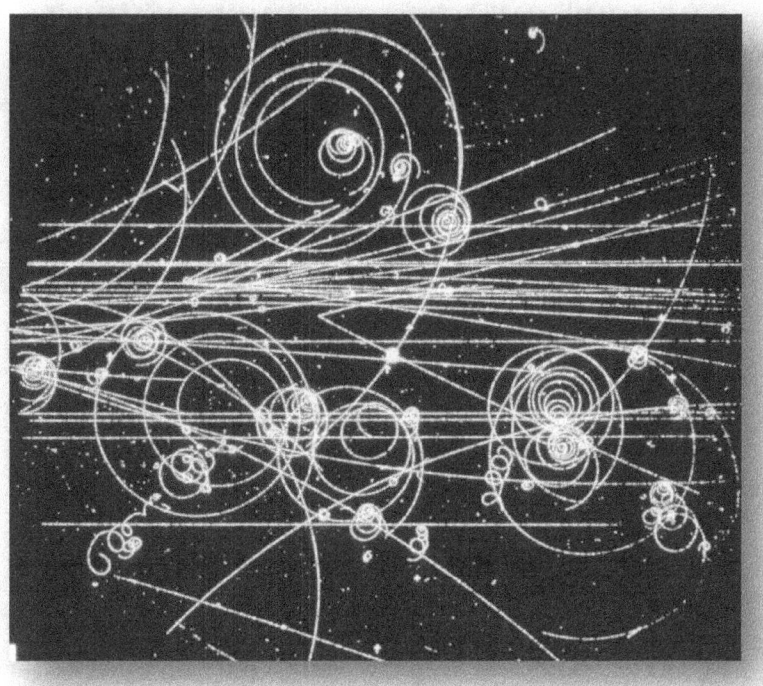

Andrew Zimmerman Jones does a brilliant job of explaining the double slit experiment and why it is so mysterious:

"It became possible to have a light source that was set up so that it emitted one photon at a time. This would be, literally, like hurling microscopic ball bearings through the slits. By setting up a screen that was sensitive enough to detect a single photon, you could determine whether there were or were not interference patterns in this case.

One way to do this is to have a sensitive film set up and run the experiment over a period of time, then look at the film to see what the pattern of light on the screen is. Just such an experiment was performed and, in fact, it matched Young's version identically - alternating light and dark bands, seemingly resulting from wave interference.

This result both confirms and bewilders the wave theory. In this case, photons are being emitted individually. There is literally no way for wave interference to take place, because each photon can only go through a single slit at a time. But the wave interference is observed. How is this possible? Well, the attempt to answer that question has spawned many intriguing interpretations of quantum physics, from the Copenhagen interpretation to the many-worlds interpretation.

Now assume that you conduct the same experiment, with one change. You place a detector that can tell whether or not the photon passes through a given slit. If we know the photon passes through one slit, then it cannot pass through the other slit to interfere with itself.

It turns out that when you add the detector, the bands disappear! You perform the exact same experiment, but only add a simple measurement at an earlier phase, and the result of the experiment changes drastically.

Something about the act of measuring which slit is used removed the wave element completely. At this point, the photons acted exactly as we'd expect a particle to behave. The very uncertainty in position is related, somehow, to the manifestation of wave effects."

As we will see later on, how one interprets this experiment will have deep philosophical repercussions.

9 | Spooky Physics

What determines light as a wave or a particle is dependent (literally) on our measuring device. And even then we cannot know both the momentum and position of that particle/wave with absolute precision. For instance, the more we know about the electron's position, the less we know about its momentum, and vice versa.

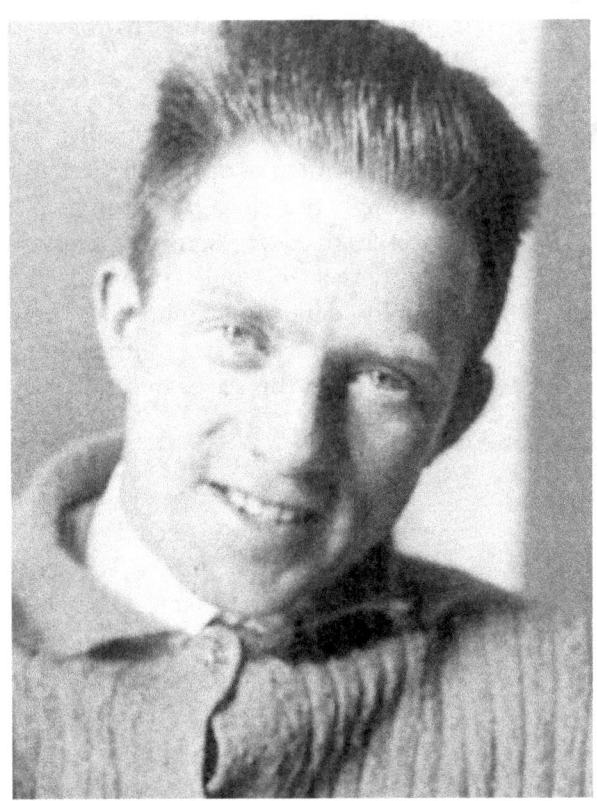

Werner Heisenberg, whose name is forever attached with this discovery (thus we have "Heisenberg's uncertainty principle") realized to the ultimate consternation of Einstein that this indeterminism forced physicists to take a different approach when trying to resolve the quanta's secrets. Instead of an exact causal explanation of any singular electron dance, what was needed was a probabilistic model of how the electron or any subatomic material behaved. It is as if nature itself was a gambling device and what it paid out was determined by odds.

The glitch, of course, for any would-be gambler (or should we say any would-be quantum mechanic) is that he or she never actually knows in one throw of the dice or one pull of the slot lever whether one is going to win or lose.

How to get around this impasse? How to beat the odds? Well, you can't actually in one throw or one pull, but you can if you gamble enough.

For instance, take a coin toss. It is either going to be heads or tails, but never both if you let it land on the ground (and not, in the very rare occasion, on its side). The odds are 50/50 which can also be translated as "I don't know." That it will be one or the other allows you some ultimate outcome that you can guess but never absolutely know in one isolated toss of the coin.

However, if you are allowed to toss the coin many times (a thousand or more times, for instance), then something else comes into play: probability functions. That is, the more you toss the coin the more you start to see patterns emerge which, given the science of statistics, will provide you with a fairly accurate gauge of what you can expect to see if one is allowed to toss a coin a million times.

A good example of this is drawn right from Las Vegas casinos (hopefully, your car has reached there by now without too much interference). The casino does not know whether or not you are going to win at the poker slot machine when you put in your dollar. Indeed, if the gambling establishment is genuine and not rigged, then it CANNOT know precisely. However, since the casino makes significant amounts of money the question arises very simply: How can that be possible if it is due to chance only? The answer is both simple and profound. Yes, the casino does not know in advance what "dice" the gambler is going to throw down on the craps table in any one isolated event, but it does have a very good understanding of the probable odds of how many winners and losers it will get if the game is played enough. This is, of course, the science of statistics.

Einstein vs. Bohr /

For instance, if you toss a quarter, there is a 50/50 chance you will get heads or tails. The odds are evenly split. However, if you toss that same coin say 1000 times, you will start to see a certain pattern emerge. You will quickly notice how difficult it is to get 200 heads in a row or 500 tails in a row. You will start to be able to calculate the odds of certain patterns emerging or not emerging. These odds, or mathematical probabilities, will start to give you some "certainty" even when dealing with something fundamentally uncertain.

So, if a friend of mine wants to bet me that she can get a 100 heads in a row, the next question I must ask her is "how many times are you going to throw it.?" Because the number of throws will either increase or lessen my confidence in taking up her bet. If she says, ah, give me a thousand tries, I would easily take her up on the bet. If, however, she starts to talk about a trillion times a trillion times, I wouldn't venture such a wager.

Quantum mechanics is essentially a probabilistic model to provide us with how an uncertain realm can yield quite predictable, even if occasionally quite odd, outcomes and trajectories.

This leads, however, to all sorts of strange and unusual paradoxes. A recent cover story on quantum theory in the *New Scientist* by Gregory T. Huang has posited four very famous illustrations of quantum weirdness:

Schrödinger's cat

Conventional quantum theory says that particles can be in a superposition of two states at once. This leads to the thought experiment of a cat being both alive and dead inside a box, depending on the state of a toxic subatomic particle. Only when you open the box or make a measurement is the animal's fate determined.

Spooky action at a distance

Einstein decried the idea of entanglement - that one particle could instantaneously affect another's spin, say, through a weird quantum link. This phenomenon, also known as non-locality, has since been demonstrated and is a key principle behind quantum computers and communications.

Objective reality

Does the moon exist if nobody is looking at it? Conventional quantum theory says there is no reality beyond what we observe, so in principle things don't exist unless they are being measured.

Uncertainty principle

If you measure the position of a quantum particle, you can't know its momentum precisely, and vice versa. The conventional explanation is that there is randomness inherent in the quantum universe.

Einstein realized that quantum theory gave astounding results and predictions, but he never felt comfortable with it as a final theory. He felt something was amiss and that at best quantum theory was an interregnum theory and that in time realism and not indeterminism would reign once again.

Einstein Doesn't Play Dice

"I think that a 'particle' must have a separate reality independent of the measurements. That is an electron has spin, location and so forth even when it is not being measured. I like to think that the moon is there even if I am not looking at it."

"Thus the last and most successful creation of theoretical physics, namely quantum mechanics (QM), differs fundamentally from both Newton's mechanics, and Maxwell's e-m field. For the quantities which figure in QM's laws make no claim to describe physical reality itself, but only probabilities of the occurrence of a physical reality that we have in view."

"I cannot but confess that I attach only a transitory importance to this interpretation. I still believe in the possibility of a model of reality - that is to say, of a theory which represents things themselves and not merely the probability of their occurrence. On the other hand, it seems to me certain that we must give up the idea of complete localization of the particle in a theoretical model. This seems to me the permanent upshot of Heisenberg's principle of uncertainty."

--Albert Einstein

What is it about quantum theory that so troubled Einstein that he would spend nearly a quarter of his life trying to find a replacement for it?

The answer is perhaps a bit simpler than we might suspect. Einstein was a realist and believed in an objective universe that exists outside of our subjective observations of it. What so bothered Einstein about quantum theory (even though he contributed to it with his photoelectric effect and Brownian motion papers and appreciated its many strengths) was that it was inherently probabilistic and that at its philosophic and methodological core was an uncertainty principle which pointed to the variability of human measurement. As John Wheeler, the eminent physicist at Cornell and Princeton and the University of Texas at Austin, later stated, "There is no phenomena unless it is an observed phenomena."

This was intolerable to Einstein since as he suggested to his eventual biographer and physics colleague, Abraham Pais, the moon really does exist even when I don't look at it.

Einstein's objections to quantum theory took two major turns. First, almost from the outset, Einstein attempted to show how the new quantum mechanics as defined by Heisenberg and Bohr was mistaken. Later, Einstein accepted to some measure the correctness of quantum theory, but tried to point out how it was an incomplete theory and most likely a bridge theory to something much more comprehensive and complete.

One of the key sticking points for Einstein was the breakdown of individual causality inherent in quantum theory, where a measuring device a priori determines the outcome of a quantum state. As Joshua Roebke in "The Reality Tests" points out,

"Schrodinger and Heisenberg independently uncovered dual descriptions of particles and atoms. Later, the theories proved equivalent. Then in 1926 Heisenberg's previous advisor, Max Born, discovered why no one had found a physical interpretation for Schrodinger's wave function. They are not physical waves at all; rather the wave function includes all the possible states of a system. Before a measurement those states exist in superposition, wherein every possible outcome is described at the same time. Superposition is one of the defining qualities of quantum mechanics and implies that individual events cannot be predicted; only the probability of an experimental outcome can be derived. (Seed, volume 16)."

The fact that quantum theory involves a connection between a measuring device and how we can ascertain reality was, for Einstein, fundamentally problematic. In a famous letter to Max Born, dated March 3, 1947, Einstein outlines why:

"I cannot make a case for my attitude in physics which you would consider at all reasonable. I admit, of course, that there is a considerable amount of validity in the statistical approach which you were the first to recognize clearly as necessary given the framework of the existing formalism. I cannot seriously believe in it because the theory cannot be reconciled with the idea that physics should represent a reality in time and space, free from spooky actions at a distance. I am, however, not yet firmly convinced that it can really be achieved with a continuous field theory, although I have discovered a possible way of doing this which so far seems quite reasonable. The calculation difficulties are so great that I will be biting the dust long before I myself can be fully convinced of it. But I am quite convinced that someone will eventually come up with a theory whose objects, connected by laws, are not probabilities but considered facts, as used to be taken for granted until quite recently. I cannot, however, base this conviction on logical reasons, but can only produce my little finger as witness, that is, I offer no authority which would be able to command any kind of respect outside of my own hand."

Perhaps the key line in the above referenced letter by Einstein is this: "I cannot seriously believe in it because the theory cannot be reconciled with the idea that physics should represent a reality in space and time, free from spooky actions at a distance."

What reality was Einstein presupposing here? An external world freed from human measurement—a world which exists truly and clearly apart from human subjectivity. But, as Einstein rightly surmised, this objective world collapses with Heisenberg's uncertainty principle, since external reality at its

13 | Spooky Physics

most fundamental constituency (atoms) is absolutely unknowable, except through a measuring device which in and of itself alters what is known. In other words, quantum mechanics is a statement about reality itself and what it is saying is that there is no world "out there" apart from our observations of it. Our observations, in other words, are part and parcel of what we observe. The dualistic idea of a world apart from our selves is a fiction. For Einstein this was the very antithesis of science in general and physics in particular. The whole scientific enterprise was predicated on the notion of an external world which was independent of the machinations of the subjective participants that arose within it.

But the real culprit here in Einstein's mind is the introduction of probability and statistics as a final pathway for understanding the underlying laws of subatomic materials. While Einstein readily concedes the powerful utility of Born's statistical understanding of wave matrices, his "little finger" tells him that quantum mechanics is merely a prelude to a greater and more unified theory which will eventually transcend probability functions and yield a straightforward and causal and objective explanation of how and why matter behaves the way it does.

As Einstein near the end of his life pointed out, "It seems to be clear, therefore, that Born's statistical interpretation of quantum theory is the only possible one. The wave function does not in any way describe a state which could be that of a single system; it relates rather to many systems, to an 'ensemble of systems' in the sense of statistical mechanics."

Further he elaborates on why he finds the statistical method a transitory one:

"Thus the last and most successful creation of theoretical physics, namely quantum mechanics (QM), differs fundamentally from both Newton's mechanics, and Maxwell's e-m field. For the quantities which figure in QM's laws make no claim to describe physical reality itself, but only probabilities of the occurrence of a physical reality that we have in view.... I cannot but confess that I attach only a transitory importance to this interpretation. I still believe in the possibility of a model of reality - that is to say, of a theory which represents things themselves and not merely the probability of their occurrence. On the other hand, it seems to me certain that we must give up the idea of complete localization of the particle in a theoretical model. This seems to me the permanent upshot of Heisenberg's principle of uncertainty."

Einstein vs. Bohr /

Why was Einstein so recalcitrant to a theory which measured only probabilities, especially if those very probabilities led to amazingly exact results? Some scholars have suggested that Einstein stubbornness was due to his personal psychology which looked for an order that he didn't see in the world of human affairs. Or, perhaps it stemmed from Einstein's first epiphany as a young boy at the age of eleven where he was able to prove for himself Pythagoras' theorem.

Along this line of reasoning, it has been argued that Einstein's passion in science was fueled by his even greater passion for discovering a truth apart from human artifice. In any case, whatever personal motivations lie behind Einstein's resistance to a purely statistical interpretation of physics, it is unassailable that he also found it philosophical objectionable. One of Einstein's more pregnant, even if cryptic, remarks about human ideas and reality is captured in his January 27th 1921 lecture to the Prussian Academy of Sciences in Berlin, Germany:

"At this point an enigma presents itself which in all ages has agitated inquiring minds. How can it be that mathematics, being after all a product of human thought which is independent of experience, is so admirably appropriate to the objects of reality? Is human reason, then, without experience, merely by taking thought, able to fathom the properties of real things. In my opinion the answer to this question is, briefly, this:—As far as the laws of mathematics refer to reality, they are not certain; and as far as they are certain, they do not refer to reality."

There are many ways to interpret what Einstein actually means here, especially in light of its philosophic import. But I think it presents a clearer beacon into why Einstein would have resisted a purely mathematical interpretation of physics, as was presented several years later by Born, Heisenberg, Bohr, et. al., in their formulation of quantum mechanics.

Einstein, ever being the realist, understood that human concepts were in themselves limited in their import and thus to conflate a theory in its present state for the ultimate state of reality was not only mistaken but wholly naïve. Ironically, in this sense, Einstein was a metaphysician whose "little finger" or "intuition" pointed beyond mere empiricism.

But Einstein's metaphysic wasn't of a religious or a spiritual kind, but rather for a reality that literally transcends human cognition and which forever escapes human thought to entrap it. As Einstein explained in his lengthy analysis of Bertrand Russell's theory of knowledge:

"In the evolution of philosophical thought through the centuries the following question has played a major role: what knowledge is pure thought able to supply independently of sense perception? Is there any such knowledge? If not, what precisely is the relation between our knowledge and the raw material furnished by sense impressions?

There has been an increasing skepticism concerning every attempt by means of pure thought to learn something about the 'objective world', about the world of 'things' in contrast to the world of 'concepts and ideas'. During philosophy's childhood it was rather generally believed that it is possible to find everything which can be known by means of mere reflection. It was an illusion which anyone can easily understand if, for a moment, he dismisses what he has learned from later philosophy and from natural science; he will not be surprised to find that Plato ascribed a higher reality to 'ideas' than to empirically experienceable things. Even in Spinoza and as late as in Hegel this prejudice was the vitalising force which seems still to have played the major role.

The more aristocratic illusion concerning the unlimited penetrative power of thought has as its counterpart the more plebeian illusion of naive realism, according to which things 'are' as they are perceived by us through our senses. This illusion dominates the daily life of men and of animals; it is also the point of departure in all of the sciences, especially of the natural sciences."

As Russell wrote;

'We all start from naive realism, i.e., the doctrine that things are what they seem. We think that grass is green, that stones are hard, and that snow is cold. But physics assures us that the greenness of grass, the hardness of stones, and the coldness of snow are not the greenness, hardness, and coldness that we know in our own experience, but something very different. The observer, when he seems to himself to be observing a stone, is really, if physics is to be believed, observing the effects of the stone upon himself.'

Gradually the conviction gained recognition that all knowledge about things is exclusively a working-over of the raw material furnished by the senses. Galileo and Hume first upheld this principle with full clarity and decisiveness. Hume saw that concepts which we must regard as essential, such as, for example, causal connection, cannot be gained from material given to us by the senses. This insight led him to a skeptical attitude as concerns knowledge of any kind.

Man has an intense desire for assured knowledge. That is why Hume's clear message seemed crushing: the sensory raw material, the only source of our knowledge, through habit may lead us to belief and expectation but not to the knowledge and still less to the understanding of lawful relations.

Then Kant took the stage with an idea which, though certainly untenable in the form in which he put it, signified a step towards the solution of Hume's dilemma: whatever in knowledge is of empirical origin is never certain. If, therefore, we have definitely assured knowledge, it must be grounded in reason itself. This is held to be the case, for example, in the propositions of geometry and the principles of causality.

These and certain other types of knowledge are, so to speak, a part of the implements of thinking and therefore do not previously have to be gained from sense data (i.e. they are a priori knowledge).

Today everyone knows, of course, that the mentioned concepts contain nothing of the certainty, of the inherent necessity, which Kant had attributed to them. The following, however, appears to me to be correct in Kant's statement of the problem: in thinking we use with a certain right, concepts to which there is no access from the materials of sensory experience, if the situation is viewed from the logical point of view. As a matter of fact, I am convinced that even much more is to be asserted: the concepts which arise in our thought and in our linguistic expressions are all- when viewed logically- the free creations of thought which cannot inductively be gained from sense experiences. This is not so easily noticed only because we have the habit of combining certain concepts and conceptual relations (propositions) so definitely with certain sense experiences that we do not become conscious of the gulf- logically unbridgeable- which separates the world of sensory experiences from the world of concepts and propositions. Thus, for example, the series of integers is obviously an invention of the human mind, a self-created tool which simplifies the ordering of certain sensory experiences. But there is no way in which this concept could be made to grow, as it were, directly out of sense experiences.

As soon as one is at home in Hume's critique one is easily led to believe that all those concepts and propositions which cannot be deduced from the sensory raw material are, on account of their 'metaphysical' character, to be removed from thinking. For all thought acquires material content only through its relationship with that sensory material. This latter proposition I take to be entirely true; but I hold the prescription for thinking which is grounded on this proposition to be false. For this claim- if only carried through consistently- absolutely excludes thinking of any kind as 'metaphysical'.

In order that thinking might not degenerate into 'metaphysics', or into empty talk, it is only necessary that enough propositions of the conceptual system be firmly enough connected with sensory experiences and that the conceptual system, in view of its task of ordering and surveying sense experience, should show as much unity and parsimony as possible. Beyond that, however, the 'system' is (as regards logic) a free play with symbols according to (logically) arbitrarily given rules of the game. All this applies as much (and in the same manner) to the thinking in daily life as to the more consciously and systematically constructed thinking in the sciences.

By his clear critique Hume did not only advance philosophy in a decisive way but also- though through no fault of his- created a danger for philosophy in that, following his critique, a fateful 'fear of metaphysics' arose which has come to be a malady of contemporary empiricist philosophising; this malady is the counterpart to that earlier philosophising in the clouds, which thought it could neglect and dispense with what was given by the senses. ... It finally turns out that one can, after all, not get along without metaphysics."

17 | Spooky Physics

In summary the reason Einstein so resisted the philosophical implications of quantum theory (the observer alters the observed) was because it puts the cart before the horse, or, more accurately in this context, it puts man's present understanding prior to the world itself. And that world, unlike man's changing views of it, isn't subjected to the whims of current scientific theory. Perhaps this is why Einstein resisted the vast majority of his colleagues who accepted the idea that what quantum mechanics presented was the limits of what could ever be known. Einstein's underlying metaphysic was that science was an attempt to bypass man's limited understanding over time and hence to make an interregnum theory final was to ignore both history and reality.

As Einstein so famously stated, "Quantum theory is certainly imposing. But an inner voice tells me that it is not yet the real thing. Quantum theory says a lot, but does not really bring us any closer to the secret of the Old One. I, at any rate, am convinced that He (God) does not throw dice."

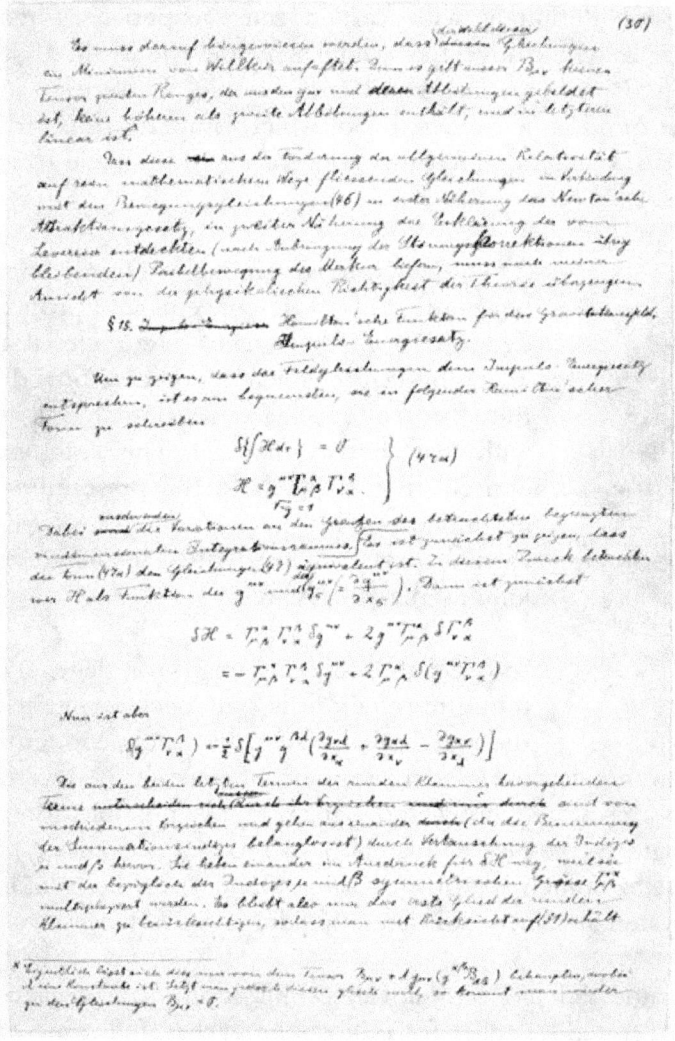

Einstein vs. Bohr /

Bohr Plays Poker

"The great extension of our experience in recent years has brought light to the insufficiency of our simple mechanical conceptions and, as a consequence, has shaken the foundation on which the customary interpretation of observation was based."

"Physics is to be regarded not so much as the study of something a priori given, but rather as the development of methods of ordering and surveying human experience." –Niels Bohr

Niels Bohr received his Nobel Prize in physics in 1922, a year after Albert Einstein's award in 1921, though both were given their awards at the same ceremony in 1922 in Stockholm. Einstein and Bohr had a deep fondness and respect for each other and while they certainly had their philosophic disagreements over the years, particularly over how to interpret the new physics, their admiration for each other lasted till the end of their lives.

It has been mentioned in several books dealing with the Einstein-Bohr debate that Einstein was more of a realist when it came to science and Bohr was more of an idealist. This description of their differences is too simplistic to be accurate.

Niels Bohr was deeply involved from the very beginning with the revolution which took place in physics during the first quarter or so of the 20th century. Indeed, his early model of the atom, based in part upon Ernest Rutherford's investigations, was an elemental bridge to later quantum theories which eventually made it obsolete. It was because of Bohr's simple, but predictive, explanation of the spectral lines of the hydrogen atom that significant progress was made in unearthing the inner workings of physical constants at the subatomic realm.

Bohr, unlike Einstein, enjoyed working with a series of devoted students and loved the to and fro of debating the implications of the latest findings in atomic theory.

It has been convincingly argued by Donald Murdoch in his groundbreaking study, *Niel's Bohr's Philosophy of Physics*, that Bohr was less an idealist and more a pragmatist when it came to interpreting the implications of quantum mechanics. What this means is that Bohr tried to let the physics itself lead to its own interpretation and not try to impose upon it his own already made philosophy.

Spooky Physics

This is best captured in one of his most famous quotes, where Bohr ruminates, "When it comes to atoms, language can be used only as in poetry. The poet, too, is not nearly so concerned with describing facts as with creating images. It is wrong to think that the task of physics is to find out how Nature is. Physics concerns what we say about Nature."

What Bohr reveals here is a deep understanding of the very limits of the scientific enterprise and how human investigations of objective phenomena are intimately limited by its own apparatus. This raises a philosophic conundrum which is age-old and is perhaps best articulated by Immanuel Kant.

As the website *Philosophy Pages* illuminates:

"According to Kant, it is vital always to distinguish between the distinct realms of phenomena and noumena. Phenomena are the appearances, which constitute the our experience; noumena are the (presumed) things themselves, which constitute reality. All of our synthetic a priori judgments apply only to the phenomenal realm, not the noumenal. (It is only at this level, with respect to what we can experience, that we are justified in imposing the structure of our concepts onto the objects of our knowledge.) Since the thing in itself (Ding an sich) would by definition be entirely independent of our experience of it, we are utterly ignorant of the noumenal realm.

Thus, on Kant's view, the most fundamental laws of nature, like the truths of mathematics, are knowable precisely because they make no effort to describe the world as it really is but rather prescribe the structure of the world as we experience it. By applying the pure forms of sensible intuition and the pure concepts of the understanding, we achieve a systematic view of the phenomenal realm but learn nothing of the noumenal realm. Math and science are certainly true of the phenomena; only metaphysics claims to instruct us about the noumena."

To grapple with quantum indeterminacy, Bohr developed his idea of Complementarity to help explain one of the chief aspects of how nature reveals itself. And because nature is embedded with complementarity, it is nay impossible to exorcise it away from scientific investigations. In fact, Heisenberg's principle of uncertainty is a defining example of how nature is paired and manifests in ways similar to the Taoist notion of Yin and Yang, or in this case, wave and particle.

As the *Wikipedia* entry on *Complementarity* elaborates: "A profound aspect of Complementarity is that it not only applies to measurability or knowability of some property of a physical entity, but more importantly it applies to the limitations of that physical entity's very manifestation of the property in the physical world. All properties of physical entities exist only in pairs, which Bohr described as complementary or conjugate pairs (-which are also Fourier transform pairs). Physical reality is determined and defined by manifestations of properties which are limited by trade-offs between these complementary pairs. For example, an electron can manifest a greater and greater accuracy of its position only in even trade for a complementary loss in accuracy of manifesting its momentum.

This means that there is a limitation on the precision with which an electron can possess (i.e., manifest) position, since an infinitely precise position would dictate that its manifested momentum would be infinitely imprecise, or undefined (i.e., non-manifest or not possessed), which is not possible. The ultimate limitations in precision of property manifestations are quantified by the Heisenberg uncertainty principle and Planck units. Complementarity and Uncertainty dictate that all properties and actions in the physical world are therefore non-deterministic to some degree." Bohr's overall view dovetails with Ernest Mach's and represents a form of logical positivism. As Jan Faye states,

"Bohr's idea of complementarity thus understood was not so different from Neurath's and Carnap's view of relating all statements about theoretical entities to statements about observable things expressed in terms of protocol sentences. Against Einstein's metaphysical attitude towards a physical reality consisting of things-in-themselves, Bohr could just reply that it does not make sense to operate with a conception of reality other than one which can be described in sentences concerning our empirical knowledge. If experimental knowledge does prohibit an ascription of a precise position and a precise momentum at the same time, it does not make sense to talk about a free, undisturbed electron to have such values anyhow."

It in this sense that Dugald Murdoch sees Bohr's philosophy as pragmatic and not preset. Whereas Einstein would follow his intuition about how nature must or should work, Bohr argued for following the data and letting it determine whatever philosophical course would follow. This is wittingly captured with Bohr's reaction to Einstein's famous dictum that God doesn't play dice when he pronounced, "Einstein, don't tell God what to do."

It can also be that because Bohr worked so closely with those who developed quantum mechanics, specifically his star pupil Heisenberg, that he was more acquainted in a practical way with what worked and what didn't. Bohr got his hands dirty with quantum theory perhaps in a way that Einstein didn't. And due to that was more willing to allow for its radical implications.

As Bohr warned, "Those who are not shocked when they first come across quantum mechanics cannot possibly have understood it." Bohr became the champion of the single most popular philosophic interpretation of the new physics, which would later be known as the Copenhagen interpretation because of the location of his institute.

In many ways, Bohr's reasoning is akin to what we find in Plato's allegory of the cave, as found in his *Republic*, but with one very telling caveat. In Plato's story we learn that prisoners shackled in the cave cannot actually see the light itself which is casting the varying shadows on the wall. And only later when unhinged can they progress from the rudimentary impressions to clearer shapes and outlines until the full luminosity of the light explains more fully how all these images were generated.

In the quantum mechanical world we are in a similar position, since we cannot actually know both the position and the momentum of any single electron, but only its probabilities and even then how we measure such an outcome predetermines its wave or particle manifestation. What the electron is "really" doing nobody knows.

Apparently nobody can know what a single bit of matter is ultimately doing, since even that definition of "bit" of matter is itself a construct, a theoretical map in order to make sense out of one aspect of what appears at such minute levels of matter. What we get when we penetrate the subatomic realm isn't, to quote Kant, the thing in itself, but only what appears visible to our intervening devices. And since we cannot intrude into that realm without some type of instrument (even a single photon cascading off an electron causes a disruption of the assumed virgin state), we don't unlock nature pure and pristine, but as nature reacts to our measuring devices. In other words, we cannot unlock nature as nature, or electron as electron, or matter as matter, since we are invariably altering what we are examining.

We might occasionally acknowledge this interference even at the macroscopic level (sociologists and psychologists are well versed in interpreter's biases in grappling with raw data), but at the quantum level it looms so large and is so evidential that its impact cannot at any instance be ignored.

Heisenberg's principle of uncertainty isn't merely a temporary limit to man's knowledge, according to Bohr, but a fundamental statement about what that knowledge is. It is for this reason that Plato's allegory is instructive, since we are not in the position of the narrator to look objectively upon the cave from the outside and the inside simultaneously. Rather, we are the prisoners in the cave and only from that position can we both induce and deduce what may or may not be ultimately real, but in so doing we are still at the Kantian level of phenomena.

What quantum mechanics revealed was precisely this epistemological impasse and how it plays out in trying to form a picture about reality. Reality we can never know, since that very concept is itself a fiction which implies that we can somehow act as an objective narrator to the entire cosmos, with a 360 degree purview and a level of certainty which implies that we are impartial witnesses to a play with a beginning, middle, and an end.

No, we are literally like the prisoners in Plato's allegory of the cave, limited by our very existence in what can and cannot know. For Bohr this wasn't merely a philosophical extension of his Kierkegaardian leanings, but the very result of what quantum mechanics revealed about our ability to come to grips with nature and how it responds to our introspections. As Bohr put it, "It is wrong to think that the task of physics is to find out how Nature is. Physics concerns what we say about Nature."

Or, as Bohr himself discovered,

"For a parallel to the lesson of atomic theory regarding the limited applicability of such customary idealisations, we must in fact turn to quite other branches of science, such as psychology, or even to that kind of epistemological problems with which already thinkers like Buddha and Lao Tsu have been confronted, when trying to harmonize our position as spectators and actors in the great drama of existence Everything we call real is made of things that cannot be regarded as real."

It is little wonder, therefore, that so many eminent scientists have had such ambivalent reactions and feelings to the implications of quantum mechanics. This is epitomized by a close reading of the following quotes garnered from the Quantum World website:

Feynman

Quantum mechanics is magic. *Daniel Greenberger*.

Those who are not shocked when they first come across quantum theory cannot possibly have understood it. *Niels Bohr*.

If you are not completely confused by quantum mechanics, you do not understand it. *John Wheeler*.

It is safe to say that nobody understands quantum mechanics. *Richard Feynman*.

If [quantum theory] is correct, it signifies the end of physics as a science. *Albert Einstein*.

I do not like [quantum mechanics], and I am sorry I ever had anything to do with it. *Erwin Schrödinger*.

Quantum mechanics makes absolutely no sense. *Roger Penrose*.

Spooky Physics

The Einstein-Bohr Crapshoot

Whereas Einstein didn't believe in a God that plays dice in the universe, Bohr not only accepted such indeterminacy but pointed out that it was part and parcel of how we understand the world of physics. Interestingly, Bohr not only acknowledged the cosmic crapshoot, but pointed out that such a game was played in the dark and it was only when we shined some light on the proceedings that we could determine its present outcome. Ironically, our very act of illuminating the hidden play fundamentally alters what we unearth.

It is as if God is playing poker in the dark and we cannot see what hand he is holding until we turn on the lights. But that very act of turning on that light can in and of itself change a face card to a number card or vice versa. Nature is like a very fine and delicate Swiss watch with many extraordinarily small and complicated and interlocking pieces hidden behind a silver chamber. We are like a brutish man with very large hands whose fingers lack any finesse or dexterity trying to figure out exactly how that watch works. But every time we try to understand its sophisticated mechanism we invariably mangle its parts by our clumsiness. Thus our very act of trying to understand or fix the watch changes, to some degree, its constituent parts.

It is for this reason that Bohr could say with confidence that we don't see nature as nature, but as nature is revealed to us through our acts of measurement, which may be more accurately described as acts of intrusion.

Both Bohr and Einstein were troubled by the new physics and the decades long discussion/debate they carried on over the implications of quantum theory provides us with one of the great philosophical debates of the 20th century.

Some commentators have outlined the Einstein-Bohr debate into four stages, starting with the Solvay Conference of 1927. Others have suggested that the debate took two major developments. While still others have argued that it was rather just one long debate which evolved over time.

Einstein vs. Bohr /

Regardless of how the Einstein-Bohr debate is partitioned, it is widely accepted that the discussion got its first fireworks at the Fifth Conference of Physics at Solvay when Einstein strenuously objected to quantum indeterminacy.

Einstein ingenuously came up with thought experiments which tried to show how uncertainty relations could be overcome and thus violate the notion of indeterminacy. At first Einstein's critique was predicated upon a modification of the famous double-slit light experiment, where he suggested that some form of measurement, albeit merely theoretical and infinitesimally small, could indeed be made which would violate the notion of indeterminism.

At first, it looked as if Einstein had provided a penetrating body blow to the new physics, but Niels Bohr brilliantly demonstrated that even in light of Einstein's updated modification it would still be impossible to gather the precision necessary to refute indeterminacy. As one commentator summarized its more technical aspects, "Bohr observes that extremely precise knowledge of any (potential) vertical motion of the screen is an essential presupposition in Einstein's argument. In fact, if its velocity in the direction X before the passage of the particle is not known with a precision substantially greater than that induced by the recoil (that is, if it were already moving vertically with an unknown and greater velocity than that which it derives as a consequence of the contact with the particle), then the determination of its motion after the passage of the particle would not give the information we seek. However, Bohr continues, an extremely precise determination of the velocity of the screen, when one applies the principle of indeterminacy, implies an inevitable imprecision of its position in the direction X. Before the process even begins, the screen would therefore occupy an indeterminate position at least to a certain extent (defined by the formalism)."

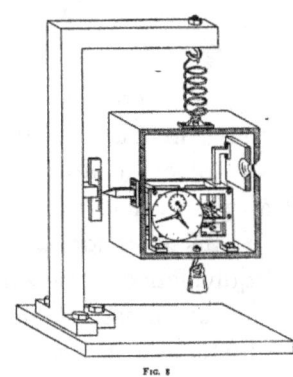

The problem that was haunting Einstein here was one of measurement, since if he could show (even theoretically) that it was possible to get a precise fix on a quanta event it would violate Heisenberg's principle of uncertainty and show prima facie that realism could be re-introduced into the new physics. In their first formal confrontation over this matter, even despite Einstein's cleverness, Bohr showed conclusively how Einstein's thought experiment was in error.

At the next Solvay Conference, however, held in 1930, Bohr had a much more difficult time overcoming what became infamously known as "Einstein's box." This thought idea is actually fairly straightforward and not difficult, even for us armchair observers, to comprehend.

Again, relating to Heisenberg's principle of uncertainty, Einstein imagined a box which contained a certain limited amount of electromagnetic radiation and which was trapped within a certain small region. Adjacent within the box was a clock which was connected to a small aperture which, given a set time, would release a photon (or small packet of radiation) from within the trapped box, thereby decreasing the amount of energy it originally contained. Connected outside of this box was a weighing scale which allowed for measuring the weight within the box before and after the photon or radiation was released. This would conceivably allow for two differing weights and thus provide one with a certainty hitherto not allowed under uncertainty relations. This thought experiment is based, in part, upon Einstein's famous equation of $E=MC^2$, where matter is literally congealed energy and thus carries weight which is amenable to some form of measurement.

Imagine the weight of Einstein's box with some bundled radiation and imagine the weight of that same box which has released through its portal a quanta of energy. It should be possible, given this scenario (which also contains a clock to accurately provide the time when that photon is released), to gather precise information about such electromagnetic energy that is not allowed under indeterminate coordinates.

In sum, Einstein's box should contradict indeterminism and thus allow for a realistic interpretation (and not merely a probabilistic one) for what transpires at the subatomic realm.

The simplicity of the experiment makes it look at first glance exceedingly convincing. Indeed, it did look to be true, even to Bohr who apparently was flummoxed when he first learned of it.

As Leon Rosenfeld commented, "It was a real shock for Bohr...who, at first, could not think of a solution. For the entire evening he was extremely agitated, and he continued passing from one scientist to another, seeking to persuade them that it could not be the case, that it would have been the end of physics if Einstein were right; but he couldn't come up with any way to resolve the paradox. I will never forget the image of the two antagonists as they left the club: Einstein, with his tall and commanding figure, who walked tranquilly, with a mildly ironic smile, and Bohr who trotted along beside him, full of excitement."

However, Bohr eventually saw the flaw in Einstein's Box, and through a crafty use of reasoning, which ironically employed using Einstein's own great discoveries against himself, he was able to show why the device wouldn't work as predicted.

As one encyclopedia entry on the subject elaborates,

"The "triumph of Bohr" consisted in his demonstrating, once again, that Einstein's subtle argument was not conclusive, but even more so in the way that he arrived at this conclusion by appealing precisely to one of the great ideas of Einstein: the principle of equivalence between gravitational mass and inertial mass. Bohr showed that, in order for Einstein's experiment to function, the box would have to be suspended on a spring in the middle of a gravitational field. In order to obtain a measurement of weight, a pointer would have to be attached to the box which corresponded with the index on a scale. After the release of a photon, weights could be added to the box to restore it to its original position and this would allow us to determine the weight. But in order to return the box to its original position, the box itself would have to be measured. The inevitable uncertainty of the position of the box translates into an uncertainty in the position of the pointer and of the determination of weight and therefore of energy. On the other hand, since the system is immersed in a gravitational field which varies with the position, according to the principle of equivalence the uncertainty in the position of the clock implies an uncertainty with respect to its measurement of time and therefore of the value of the interval Δt. A precise evaluation of this effect leads to the conclusion that the relation cannot be violated."

After the Sixth Physics Conference at Solvay, Einstein took a different line of criticism, since he apparently accepted (at least temporarily) the recalcitrant inherency of uncertainty. Rather, Einstein argued that though quantum mechanics provided much headway into the more esoteric realms of physics, it was nevertheless an incomplete theory. As Einstein explained, "I have the greatest consideration for the goals which are pursued by the physicists of the latest generation which go under the name of quantum mechanics, and I believe that this theory represents a profound level of truth, but I also believe that the restriction to laws of a statistical nature will turn out to be transitory....Without doubt quantum mechanics has grasped an important fragment of the truth and will be a paragon for all future fundamental theories, for the fact that it must be deducible as a limiting case from such foundations, just as electrostatics is deducible from Maxwell's equations of the electromagnetic field or as thermodynamics is deducible from statistical mechanics."

Perhaps the height of the Einstein-Bohr debate happened in 1935 when Einstein, along with Boris Podolsky and Nathan Rosen, published a landmark paper in *Physical Review* under the title, "Can Quantum-Mechanical Descriptions of Physical Reality Be Considered Complete?" This paper, perhaps more than any other Einstein has written, has generated the most heated debate about quantum theory. Because at the time it was written its profound implications were mostly overlooked or prematurely dismissed.

An abstract of the paper which was published in Volume 47, Issue 10 (see pages 777 to 780) of *Physical Review* is deceptively simple:

"In a complete theory there is an element corresponding to each element of reality. A sufficient condition for the reality of a physical quantity is the possibility of predicting it with certainty, without disturbing the system. In quantum mechanics in the case of two physical quantities described by non-commuting operators, the knowledge of one precludes the knowledge of the other. Then either (1) the description of reality given by the wave function in quantum mechanics is not complete or (2) these two quantities cannot have simultaneous reality. Consideration of the problem of making predictions concerning a system on the basis of measurements made on another system that had previously interacted with it leads to the result that if (1) is false then (2) is also false. One is thus led to conclude that the description of reality as given by a wave function is not complete."

It turns out to be one of the great ironies of this famous paper is that it ended up providing a very strong case for (and not against) quantum mechanics. What the paper sets out to do, more formally, is this (according to *Wikipedia's* entry on *EPR*):

"The EPR experiment yields a dichotomy. Either

1.	The result of a measurement performed on one part A of a quantum system has a non-local effect on the physical reality of another distant part B, in the sense that quantum mechanics can predict outcomes of some measurements carried out at B; or...

2.	Quantum mechanics is incomplete in the sense that some element of physical reality corresponding to B cannot be accounted for by quantum mechanics (that is, some extra variable is needed to account for it.) "

At the time that this paper was published, it was not yet known how to "test" its basic hypothesis, and thus it was attacked on more theoretical grounds or as in the case of Wolfgang Pauli discounted without due consideration.

Just months after Einstein's collaborative paper was published in 1935, Bohr published his own rejoinder (with the same title as Einstein's, "Can Quantum Mechanical Description of Physical Reality be Considered Complete") in the same *Physical Review* in Volume 48, Issue 8, pages 696-702. Although Bohr didn't provide an experiential rebuff to Einstein, he did lay out his point by point critique.

Argued Bohr:

"Such an argumentation, however, would hardly seem suited to affect the soundness of quantum-mechanical description, which is based on a coherent mathematical formalism covering automatically any procedure of measurement like that indicated. The apparent contradiction in fact discloses only an essential inadequacy of the customary viewpoint of natural philosophy for a rational account of physical phenomena of the type with which we are concerned in quantum mechanics. Indeed, the finite interaction between object and measuring agencies conditioned by the very existence of the quantum of action entails—because of the impossibility of controlling the reaction of the object on the measuring instruments if these are to serve any purpose—the necessity of a final renunciation of the classical ideal of causality and a radical revision of our attitude towards the problem of physical reality. In fact, as we shall see, a criterion of reality like that proposed by the named authors contains—however cautious its formulation may appear—an essential ambiguity when it is applied to the actual problems with which we are here concerned."

To understand what is at stake, it is perhaps important here to introduce the concept of quantum entanglement, where two electrons (each with opposite spins) are forever engaged with each other such that a decisive change of one electron's spin from upward to downward must (because of quanta superposition of two states) change the other twin's electron spin from downward to upward, and vice versa.

28 | Spooky Physics

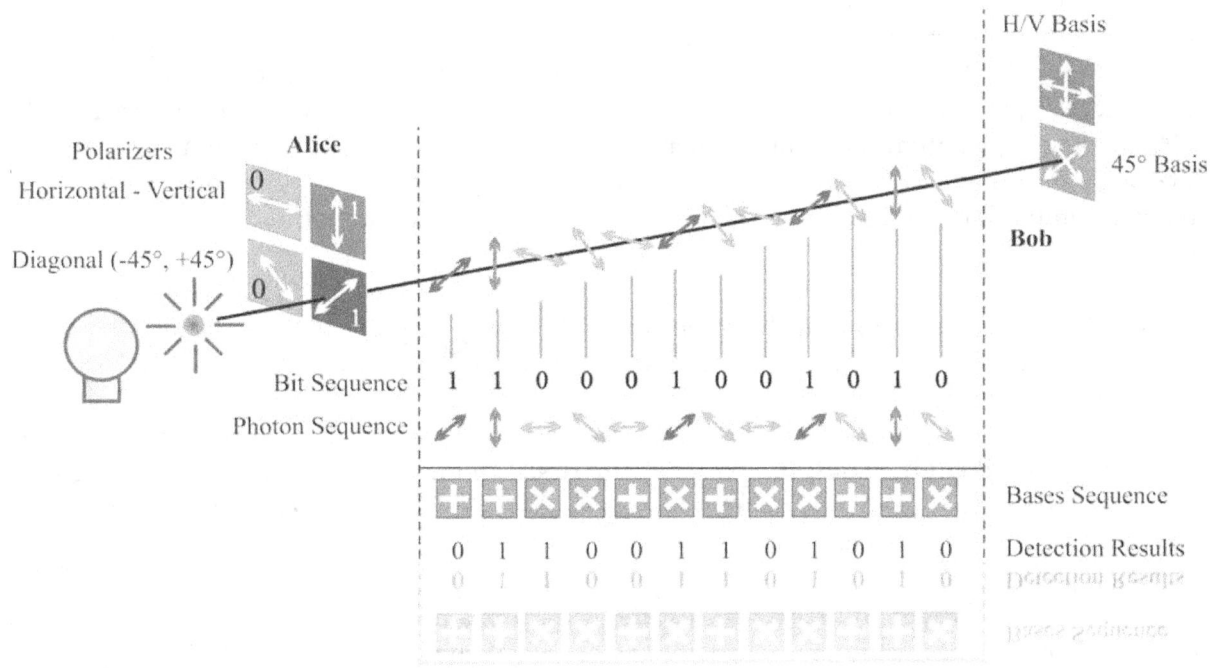

A more technical, yet precise, explanation is provided by David Bohm, J. Hilts and others. The following excerpt from an entry on quantum entanglement from the online encyclopedia *Wikipedia* appears based, at least in part, upon J. Hilts' 2007 paper in the *Journal of Physics*.

"We have a source that emits pairs of electrons, with one electron sent to destination A, where there is an observer named Alice, and another is sent to destination B, where there is an observer named Bob. According to quantum mechanics, we can arrange our source so that each emitted electron pair occupies a quantum state called a spin singlet. This can be viewed as a quantum superposition of two states, which we call state I and state II. In state I, electron A has spin pointing upward along the z-axis (+z) and electron B has spin pointing downward along the z-axis (-z). In state II, electron A has spin -z and electron B has spin +z. Therefore, it is impossible to associate either electron in the spin singlet with a state of definite spin. The electrons are thus said to be entangled.

Alice now measures the spin along the z-axis. She can obtain one of two possible outcomes: +z or -z. Suppose she gets +z. According to quantum mechanics, the quantum state of the system collapses into state I. (Different interpretations of quantum mechanics have different ways of saying this, but the basic result is the same.) The quantum state determines the probable outcomes of any measurement performed on the system. In this case, if Bob subsequently measures spin along the z-axis, he will obtain -z with 100% probability. Similarly, if Alice gets -z, Bob will get +z.

There is, of course, nothing special about our choice of the z-axis. For instance, suppose that Alice and Bob now decide to measure spin along the x-axis, according to quantum mechanics, the spin singlet state may equally well be expressed as a superposition of spin states pointing in the x direction. We'll call these states Ia and IIa. In state Ia, Alice's electron has spin +x and Bob's electron has spin -x. In state IIa, Alice's electron has spin -x and Bob's electron has spin +x. Therefore, if Alice measures +x, the system collapses into Ia, and Bob will get -x. If Alice measures -x, the system collapses into IIa, and Bob will get +x.

Einstein vs. Bohr /

In quantum mechanics, the x-spin and z-spin are "incompatible observables", which means that there is a Heisenberg uncertainty principle operating between them: a quantum state cannot possess a definite value for both variables. Suppose Alice measures the z-spin and obtains +z, so that the quantum state collapses into state I. Now, instead of measuring the z-spin as well, Bob measures the x-spin. According to quantum mechanics, when the system is in state I, Bob's x-spin measurement will have a 50% probability of producing +x and a 50% probability of -x. Furthermore, it is fundamentally impossible to predict which outcome will appear until Bob actually performs the measurement.

So how does Bob's electron know, at the same time, which way to point if Alice decides (based on information unavailable to Bob) to measure x and also how to point if Alice measures z? Using the usual Copenhagen interpretation rules that say the wave function "collapses" at the time of measurement, there must be action at a distance or the electron must know more than it is supposed to. To make the mixed part quantum and part classical descriptions of this experiment local, we have to say that the notebooks (and experimenters) are entangled and have linear combinations of + and – written in them, like Schrödinger's Cat."

As this is a fairly complicated and technical feature in quantum mechanics, varying physicists from Erwin Schrodinger (thus the famous "Schrodinger's cat") to David Bohm, have tried to explicate it by using ordinary objects that we are all familiar with.

To further illustrate what is at stake here and to perhaps underline why quantum mechanics has been described as "weird," imagine that the paired electrons are actually a deeply in love married couple far into the future. After their initial honeymoon, the couple (we will call them Brad and Angelina) have to go back to work on their respective planets (they met on an interstellar dating service over the trans-galaxy web service), which are in completely different solar systems, separated by a billion miles. Since our entangled pair, like their electron counterparts, represent the dynamic fusion of opposing spins (the female/male interplay), further imagine that if Brad was to have a sex change operation and turn himself into a she, his wife, Angelina, must (given this obviously forced analogy) in turn change herself into a "he."

The question that arises here, as it does with paired electrons, is how long would the change take and how would it be implemented? In other words, how would Angelina find out that her lover Brad has become "her" so that she may become "him"? In a conventional physics sense, we are tackling the issue of how information travels and how long it takes to traverse spatial distances. More pointedly, we are coming to grips with the very foundation of modern physics and how matter behaves. At the quantum level, however, we have discovered that things operate quite differently than we ever expected. Given the speed limit that has defined how fast objects can travel (basically

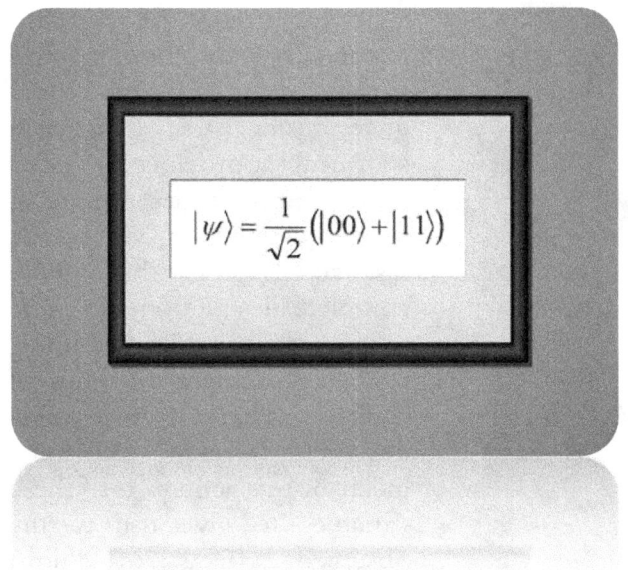

the speed of light, 186,000 plus miles per second), we would expect the information about Brad's sex change to reach Angelina in about an hour and a half, give or take a few minutes depending on initial conditions. What we would not expect is for such information to reach Angelina in no time at all.

It was in reaction to this absurd claim (something nonlocal could actually influence a very specific local event) that Einstein used his pithy phrase, "spooky actions at a distance." In his 1935 paper with Podolsky and Rosen, Einstein had no idea at the time that the very objection he was making about quantum theory was in itself the basis for a hypothetical experiment which would decades later actually be performed and show, quite conclusively, that spooky action at a distance (or nonlocal interference) was indeed part and parcel of quantum reality.

Writes Einstein:

"One could object to this conclusion [the one Einstein was making about quantum theory not being complete] on the grounds that our criterion of reality is not sufficiently restrictive. Indeed, one would not arrive at our conclusion if one insisted that two or more physical quantities can be regarded as simultaneous elements of reality only when they can simultaneously measured or predicted. On this point of view, since either one or the other, but not both simultaneously, of the quantities P and Q can be predicted, they are not simultaneously real. This makes the reality of P and Q depend upon the process of measurement carried out on the first system, which does not disturb the second system in any way. No reasonable definition of reality could be expected to permit this."

But this very last quoted line in what is known more commonly as the EPR paper (so named because of the initials of the three authors) is precisely what does happen in quantum entanglement. It is precisely what does happen when Brad gets a sex change operation on a distant planet and becomes a female and Angelina instantly turns into a man, even though she is a billion miles away. Einstein's spooky actions at a distance are right, even if he coined that phrase as a pejorative slight on the utter silliness of the notion.

At the time of this paper, however, there was no way of knowing that it would serve as the impetus for J.S. Bell to devise an experiment to find out if hidden, but local, events were really transpiring at the quantum level or, rather if quantum mechanics was indeed a complete description and something non-local was occurring. As J. Hilts wrote in his review of Einstein and Bohr's 1935 papers:

"With these results [as shown in Bohr's experiment as mentioned in his paper] Bohr claimed that the description of physical reality given by EPR was wrong. Their conclusion regarding the quantum mechanical incompleteness of the description of reality is thus also false.

Spooky Physics

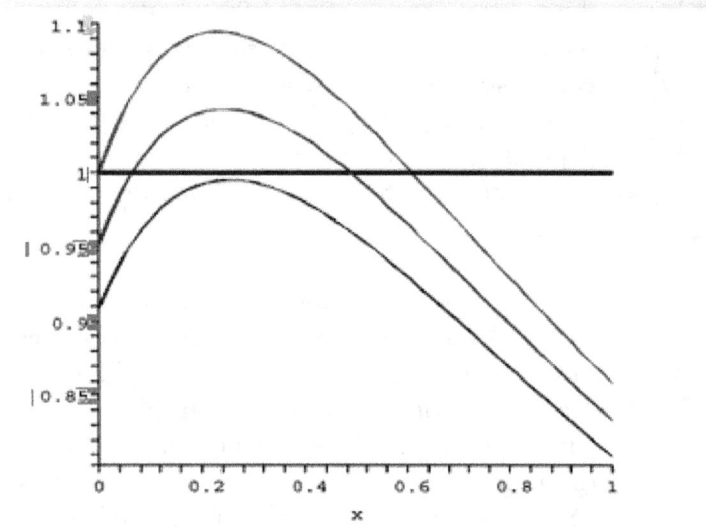

FIG. 1: *The loss of violation as decoherence increases.* The function is normalized to its maximal classical value: $C = 1$. The variable $x = |v|^2$ gives the square amplitude of the field configurations. The occupation number is $n = 100$, and the three values of δ are 0 (red, upper), the pure state, 0.05 (blue, middle), and 0.1 (brown, lower), the border regime. The function $C(v, -v, \delta)$ is asymptotically independent of n in the large n limit.

The conclusions of the EPR paper try to resolve this paradox by stating that quantum mechanics is merely a statistical approximation of a more complete description of nature which has yet to be discovered. In this more complete description of nature there exists variables pertaining to every element of physical reality. There must be, however, some unknown mechanism acting on these variables to give rise to the observed effects of "non-commuting quantum observables." Such a theory is called hidden variable theory.

John S. Bell derived a set of inequalities, known as Bell's Inequalities, which showed that the predications [sic: predictions?] of quantum mechanics through the EPR thought experiment actually differed from the predictions of various hidden variable theories. These predictions have much stronger statistical correlations between measurement results performed on different axes than the hidden variable theories. These theories are generally non-local; recall the EPR paper used locality as one of their arguments.

Today most physicists believe that the EPR "paradox" is only a paradox because our classical intuitions do not correspond to physical reality in the realm of quantum mechanics."

Although Bohr wrote a fairly lengthy critique of Einstein's position, he didn't know enough at the time of non-local variables to drive home the point that spooky action at a distance is indeed allowed and predicted by quantum theory. Indeed, if non-local influences would have been known then, Einstein couldn't have written, "No reasonable definition of reality could be expected to permit this." Yet five decades later, such a definition of reality (albeit at the quantum level) turned out to be both reasonable and true. As the CNRS website in France explains:

"In 1974, Aspect began probing the subject, building upon the pioneering work of John Clauser and collaborators. He understood how to test the locality hypothesis, central in the controversy. He developed polarizers whose settings could be changed every ten nanoseconds and set up a source of entangled photons with an unprecedented efficiency. The key experiments, carried out at Orsay in 1982 by Aspect, Philippe Grangier, Gérard Roger, and Jean Dalibard, showed a clear violation of Bell's inequalities in conditions closely resembling the ideal "Gedanken Experiment"—the foundation for the theoretical discussions. Quantum theory was once again vindicated. "A pair of entangled photons should be considered as a global, inseparable quantum system," Aspect concludes. Twenty years later, it appears this work has helped in launching the second quantum revolution, with promises for quantum cryptography and quantum information processing."

Who Won the Game?

The most interesting feature of the Einstein-Bohr debate is that even though both physicists have been dead for over nearly a half century (Einstein in 1955 and Niels Bohr in 1962), the debate they started in the 1920s is still continuing. Some physicists, such as David Bohm, have championed newer versions of realism where quantum indeterminacy is resolved by introducing such notions as the "pilot-wave" model which allows for reintroducing "actual positions" for particles "without the traditional invocation of a special, and somewhat obscure, status for observation." (The hallmark of the Copenhagen interpretation of quantum theory). As the *Stanford University Encyclopedia on Philosophy* explains:

"Bohmian mechanics, which is also called the de Broglie-Bohm theory, the pilot-wave model, and the causal interpretation of quantum mechanics, is a version of quantum theory discovered by Louis de Broglie in 1927 and rediscovered by David Bohm in 1952. It is the simplest example of what is often called a hidden variables interpretation of quantum mechanics. In Bohmian mechanics a system of particles is described in part by its wave function, evolving, as usual, according to Schrödinger's equation. However, the wave function provides only a partial description of the system. This description is completed by the specification of the actual positions of the particles. The latter evolve according to the "guiding equation," which expresses the velocities of the particles in terms of the wave function. Thus, in Bohmian mechanics the configuration of a system of particles evolves via a deterministic motion choreographed by the wave function. In particular, when a particle is sent into a two-slit apparatus, the slit through which it passes and where it arrives on the photographic plate are completely determined by its initial position and wave function.

Bohmian mechanics inherits and makes explicit the nonlocality implicit in the notion, common to just about all formulations and interpretations of quantum theory, of a wave function on the configuration space of a many-particle system. It accounts for all of the phenomena governed by nonrelativistic quantum mechanics, from spectral lines and scattering theory to superconductivity, the quantum Hall effect and quantum computing. In particular, the usual measurement postulates of quantum theory, including collapse of the wave function and probabilities given by the absolute square of probability amplitudes, emerge from an analysis of the two equations of motion — Schrödinger's equation and the guiding equation - without the traditional invocation of a special, and somewhat obscure, status for observation."

While still other physicists, such as Hugh Everett, have extended the logical implications of quantum indeterminism and postulated a many worlds hypothesis, whereby "there are myriads of worlds in the Universe in addition to the world we are aware of. In particular, every time a quantum experiment with different outcomes with non-zero probability is performed, all outcomes are obtained, each in a different world, even if we are aware only of the world with the outcome we have seen. In fact, quantum experiments take place everywhere and very often, not just in physics laboratories: even the irregular blinking of an old fluorescent bulb is a quantum experiment."

A growing number of physicists today are taking a fresh look at the philosophical implications of the Einstein-Bohr debate and suggesting that Einstein's objections to quantum theory being incomplete deserves more attention. Others have suggested that the debate can only be resolved by trying to find a grand unified theory which unites gravity with electromagnetism. Philosophically, the issue of realism in physics versus statistical approximations is a profound one and has implications for fields ranging from evolutionary psychology to Bayesian probability theories in neuroscience.

As for an ultimate winner of the Einstein-Bohr debate, it may well be that the answer to that question is as indeterminate as the position of a single photon.

> The Bohr-Einstein debate went on for decades. Einstein showed up at now-legendary European conferences with ingenious ideas that would support the idea of a real world out there. Without fail, Bohr shot down all of Einstein's arguments. But, although Bohr gradually won the allegiance of the physics community, he never convinced Einstein, who died in 1955 still believing that there must be a real world out there somewhere.
>
> -- Karl Giberson

Recommended Readings

Annotated Books on Quantum Theory

While researching the material for this monograph, *Spooky Physics*, I read several very helpful books in the field of quantum theory. These works including biographies of Albert Einstein, Niels Bohr, Erwin Schrodinger, and Max Born, as well as books on the topic of quantum physics itself. The following offers an annotated biography of just a few of my sources:

The End of the Certain World *by Nancy Greenspan*

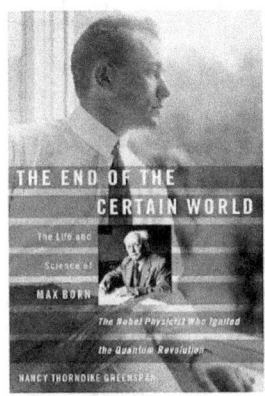

This biography of the life and times of Max Born was such a pleasure to read that I found myself a bit saddened upon its closure. It left me wanting to read even more on the life of this great physicist. Born, who inspired the world with his plea for ethical standards in science and his call to break through some of the great mysteries of the universe, was a hero of sorts. He was a man of character, tolerance and brilliance. His deep friendships with Einstein and Bohr and other renowned scientists showed the human connections he made while developing deep insight into the world of the atom.

While nurturing tight bonds with other scientists, Born's relationship with his wife, Heidi, was quite unusual. He tolerated her dalliances, especially with her eight year lover, Herglotz. Despite Heidi's romantic adventures, Born loved her and wanted to maintain a married life with her. Heidi is not necessarily an unlikable character herself. She is strong willed and insightful, and in times when Max needed direction she was there. Heidi's brief interest in Vedanta philosophy, developed while living for a stint in India, was replaced with her calling to Quaker social ideals. She remained his wife until Max's death in 1970. And when she dies two years after her husband, she was buried next to him.

It was not until the latter years of Born's life that he received the Nobel Prize in physics. For many years Max felt slighted for not receiving the prestigious award when others in his field did. Even many of the students he worked with, including Heisenberg and Pauli, were honored. But finally towards the end of his life this special award came his way. A knock on the door by a Swedish journalist announcing the news was the climax of his life. Walking down the isle, although nearly tripping in front of the king of Sweden, to receive this honor was the affirmation he so long for. On his gravesite in Gottingen, Germany reads his groundbreaking equation in quantum mechanics: $pq - qp = h/2pii$.

While Born's contribution to physics is undeniable, he himself questioned his own status in the field. When Oppenheimer omitted him when listing the great theoretical physicists of his time, Born, who once taught Oppenheimer, wrote him a letter expressing his hurt and anger. Oppenheimer's response was that he simplified the list to reduce confusion but he clearly acknowledged Born's work as the very foundation of quantum theory.

Max Born was a natural humanitarian and pacifist in similar vein with Einstein. When other scientists wanted to use their research skills for weapons research Max argued for strict ethical guidelines in science. "Love," he said, "is a power just a strong as the atom." Having to confront the Nazi world as a German Jew, though a non-religious one, was a life altering experience. Many of Max's friends and relatives were killed by the Third Reich. The stress of life during WWII took its toll on both Max and Heidi. Max suffered from bouts of illness, including severe asthma, and Heidi, who suffered from depression and exhaustion, lived months at a time in retreats.

The big debate between Einstein and Bohr about the nature of quantum mechanics was touched upon many times in this reading and was indeed the focus of my attention. Born, while extremely respectful to Einstein, argued against Einstein's position of a deterministic world, going as far as to call him "wrong." Bohr's uncertain world of quantum mechanics, though counterintuitive, he thought was an accurate understanding of nature. The "end of the certain world" is an appropriate title to grasp Born's position.
I found it interesting that the author garnered her research with the help of Olivia Newton-John, the granddaughter of Max Born. Altogether this was a remarkable read.

Niels Bohr's Times, in Physics, Philosophy, and Polity *by Abraham Paris*

The author Abraham Paris, who was good friends with Bohr, offers a brilliant biography of this scientific genius and philosopher. Bohr (died 1962 at 77 years old of heart failure) is credited with founding quantum theory. His great insight was that quantum theory violated the classic concepts of physics held sacred. Bohr's correspondence principle was his attempt to reconcile the new and old physics together. Taking Paris' lead, let us look at Bohr's politics, philosophy and, most importantly, contribution to physics.

Polity: Bohr sought an open dialogue between the West and USSR so as to prevent what everyone thought was an inevitable cold war. While his dream of openness did not come to fruition, his gallant effort in pursuit of it deserves recognition. Meeting with both Churchhill and Roosevelt to promote an open world and writing several letters to the United Nations in the 1950s on the topic resulted it little change toward post- war peace. During WWII, Bohr played a role, though minor, in the weapons program. The fear back then was that the Germans were in the race to develop atomic weapons of mass destruction. Bohr later argued that new atomic weapons could help improve international relationships as each country, armed with devastating weaponry, would take each other very seriously. Bohr wanted Russia to be consulted by Western leaders about nuclear arms in order to prevent a post-war cold war. His noble efforts went unheeded.

During WWII, Bohr helped aid refugees. He himself was under the threat of arrest by the German military police in Copenhagen and so took refuge in England. In Denmark Bohr was considered a national hero for his philanthropy and genius. He also founded the world's leading center for theoretical physics in Copenhagen and this brought world recognition to the city and country.

Philosophy: Apart from science Bohr held many other interests. He loved art and was well read in Shakespeare and in literature classics. Philosophy was certainly among his fortes.

Abraham Paris placed Bohr as one of the most notable "twentieth century philosophers." His complementarity concept applied not just to physics but to a variety of areas, including philosophy, psychology, biology and anthropology. The complementarity idea refers to "two aspects of a description that are mutually exclusive yet both necessary for a full understanding of what is to be described." This sort of reminds me of F. Scott Fitzgerald who said that the sign of an intelligent mind is the ability to hold two totally contradictory ideas at the same time and still function. While Bohr's initial concept applied to physics, specifically the idea that quanta is both wave and particle, he contended that we should take this idea of complementarity and apply it to other fields of study. Paris comments that for him the complementary way of thinking was "liberating." Interestingly, Einstein, who showed great love for Bohr, never came around to accept Bohr's way of thinking here. Einstein, instead, argued that when we look deeper we will one day see that phenomena existed independently of observation as supported by classical physics.

Moreover, in terms of philosophy, Bohr read Kierkegaard's works not just for philosophical insights (note: though baptized in the Lutheran Church, Bohr was a non religious man; he paralleled Einstein who was a non-religious figure as well), but also in admiration of his style of writing. Bohr's own philosophy seemed to parallel Kant's, specifically Kant's view that causality was not derived from experience but was an a priori judgment.

Apparently Bohr even demonstrated some interest in Eastern philosophy when he chose the Chinese symbol Yin-Yang as his emblem on his coat of arms when knighted in Denmark. This fit with his complementarity concept that opposites are indeed complementary.

Physics: Besides being considered the grandfather of nuclear medicine, Bohr is most known for being one of the key founders of quantum theory of matter. The indeterminism of quantum mechanics did not fit with the causal rules of classic physics. This Bohr full heartedly embraced along with the "epistemological lessons" it taught us, while Einstein argued that a correct understanding of quantum mechanics that reconciled old physics with the new was yet to be discovered.

Paris explains that quantum theory can be broken up into two time periods: 1900 to 1925 referred to as old quantum theory in which the science of quantum theory was established and analogies were used to understand atomic orbits; the second phase began after 1925 with the onset of quantum mechanics. Heisenberg, Born, Schrodinger, as well as Bohr, etc., mark this latter phase. Bohr's significant contributions to quantum mechanics began in 1927. While Heisenberg discovered the uncertainty principle around this time, Bohr developed the complementarity principle. This principle offers us not only a scientific understanding of wave-particle duality but also a deep philosophical insight into life.

While Einstein eventually accepted quantum mechanics, he continued to argue, unlike Bohr, that a deeper theory will one day explain what appeared to be a dichotomy between classic physics and quantum physics. Bohr's position did not waver, despite hours of intellectual debate between Einstein and Bohr. Bohr contended that no deeper theory need explain the difference between physics of the very small (quantum) and Newtonian physics. For some reason, quipped Bohr, the laws of physics break down when we went the weird world of quantum mechanics.

Thus began the famous debate between Einstein and Bohr which still has not officially been resolved. And it is not simply a physics debate but indeed a profound philosophical one as well.

A Short History of Nearly Everything *by Bill Bryson*

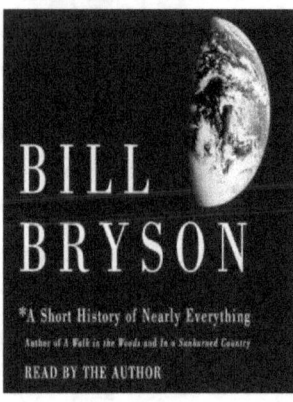

This book begins with the origins of the universe 13.7 billion years ago and then continues throughout to cover many of the major scientific advances and historical events that have made the earth that we live in today. The understanding of the atom, the discovery of the DNA structure, the extraordinary advancements we have made in geology, astronomy, anthropology are all subjects of this amazing book. If one wishes to learn a variety of scientific ideas in one read this is it. How does one cover a short history of nearly everything? Well, while this is a very difficult task, Bryson certain succeeds in familiarizing the reader with the life of Newton, Darwin, Einstein, Crick, etc., and the great advancements they each made.

Certain sections of this book caught my attention more than others. I will focus on what captured my imagination, specifically, concentrating on the sense of wonder that science undoubtedly invokes.

Astronomy: The awesomeness of our universe is a central point in the book. It is fascinating to note that at least 90% of the universe is dark matter, that which we cannot even see, and thus empty space is really not empty at all. Moreover, strangely the universe is expanding out at an accelerated rate. Scientists can actually prove that the universe is expanding by looking at what is called as "the red shift." As light moves away from us we see the red end of the light spectrum and blue as light approaches us. Through the telescope we witness red.

The Big Bang is a topic of great importance in this book. Bryson points out that one percent of the static on the TV is from the Big Bang, a moment of singularity. Perhaps we are in an eternal cycle of collapsing and expanding universes and that ours is just one of many larger universes. Physicists argue that there may be not just one universe but an infinite number of them. And ours might have no end as it folds back upon itself like a bubble ("boundless but finite").

Proxima Centauri is our nearest star and is part of a three star cluster called the Alpha Centauri. This nearest star is 100 million times farther than the moon and 4.3 light years away or 25,000 years by spacecraft. The next star would be Sirius, another 4.6 light years away. Bryson really tries to get the reader to appreciate how enormous outer space is where the "average distance between stars is 20 million million miles."

In terms of statistics, there are most likely other life forms out there but it is unrealistic, even the great distances, that we have encountered them. There are at least 100 to 400 billion stars in our Milky Way Galaxy and at least 140 billion galaxies out there. "If galaxies were frozen peas there would be enough to fill a large auditorium," states Bryson. Interestingly, he says that a conservative number puts advanced civilizations in the Milky Way in the millions. Sagan calculated that the number of possible planets in the universe is "10 billion trillion" and that if you were thrown at random in the universe the chances that you would be next to one is "one is a billion trillion trillion." There is just so much unimaginable space out there. And thank goodness for the vastness of space given that a Supernova, a star that collapses and then explodes, if nearby would destroy any life on our planet. It was about 4.5 billion years ago an object the size of Mars hit earth and the debris that came from the earth formed within a year into the moon we have today. And how fortunate we are to have this moon, for the moon's gravitational pull keeps the earth stably spinning and not wobbling off.

Bryson continues to show how amazing it is that we have life on this planet. If we were just 5% nearer to the sun or 15% farther from the sun life here could not exist.

Anthropology/Biology: Bryson also spends a great deal of effort exploring how fascinating the human being is! If one event was a bit off in the 3.8 billion years of earth's history you would not be here. One nanosecond different and there is no you. Amazing! Humans live on average 650,000 hours, a fleeting amount of time in the cosmic scheme of things. Also, most species only lasts 4 million years and 99.99 percent of all species are now extinct. What a privilege that we are here as we are now.

"Why is our fossil record so thin?" Bryson queries. Well, the chances of being fossilized are very rare. Bryson points out that "only one bone in a billion" become fossils. If this be the case then out of all of the Americans today (270-300 million) with 206 bones each only 50 bones will fossilize (1/4 of a skeleton). And then consider that we will have to find these 50 hidden fossils.

At the cellular level humans are all "youngsters." Most cells live no more than a month, and for cells that stay with you like brain cells (while you have 100 billion of them, you lose 500 of them every hour) individual components of them are also renewed monthly. Amazingly, "there isn't a single bit of us that was part of us nine years ago." Talk about reinventing ourselves.

One interesting question that Bryson tackles is: what is the genetic difference between humans? Actually, we are 99.9 % genetically the same. Four simple letters make all the diverse forms of life we see today. One time in a million there is a SNIP, a mutation. The .1% difference is due to our snips. We don't see huge mutations all around us since 97% of our DNA is junk DNA and many snips occur there. Junk DNA is still around in our code since they are good as getting copied but have no detectable consequence.

We have the same number of genes as grass (about 30,000). Sixty percent of our DNA matches that of a fruit fly. What this tells us, explains the author, is that all of life is one. Think of the awesome reality of this. Four little letters make up the ingredients for all life forms on this planet. We are all intimately connected at the deepest levels.

In this book, the author investigates our most recent ancestors. Homo erectus, it appears, is an important dividing line. Before Homo erectus the Homo species looked apelike and after looked humanlike. Early modern humans appeared to move out of Africa about 100,000 years ago. Neanderthals existed for about 100,000 years as well but died out about 35,000 years ago. It seems that there is no genetic connection between mitochondrial DNA of modern humans and Neanderthals. It is still a mystery why they died out. Perhaps we competed for the same resources, Bryson ponders. Humans have existed for only .0001% of Earth's history and in celebration, Bryson exclaims, what an "achievement" it is that we are here.

Physics: The section on physics is called a NEW AGE DAWNS. Here one learns about the beginning of the quantum age. Energy, according to Planck, can come in individual packets called quanta. It is really "liberated matter" as Einstein's $E = MC^2$ indicates. Moreover, space and time are now understood not to be absolute but relative to the one observing. The faster you go the slower time goes. Even stranger, time is part of space and is known as the dimension spacetime. Gravity can bend spacetime and warp it. Mass of any kind alters the 'fabric of the cosmos." The universe can be described as the "ultimate sagging mattress." Gravity now gets re-thought. Instead of a force it is "the byproduct of the warping of spacetime." As one physicists said, "What moves the planets and stars is the distortion of space and time."

While the Greeks first proposed atoms, it was Einstein who provided solid evidence for their existence with his 1905 paper on Brownian motion. Atoms are composed of three sections: electrons, protons and neutrons (the latter two are in the nucleus). Interestingly, "if the atom were expanded to the size of a cathedral, the nucleus" (the atom's mass; incredibly dense but only one millionth of a billionth of the total atom) "would be about the size of a fly but many thousands of times heavier than the cathedral." Most of the atom is empty space and "solidity" is really an illusion. Bryson continues: "When you sit in a chair, you are not actually sitting there, but levitating about it at a height of one angstrom (a hundred millionth of a centimeter); your electrons and its electrons implacably opposed to any closer intimacy."

The atoms that make you up are from the original stardust of the universe. Bryson points out that they have been "part of millions of organisms on the way to becoming you." Indeed, our atoms are recycled at death. At least one billion of our own atoms came from Shakespeare and from Buddha and from all the other historical greats. It takes decades for the atoms to be "redistributed." But they do go on, indefinitely, and into any variety of forms. Thus, we are reincarnated in a way.

In the world of the very small the same laws that govern the macro world do not apply. The idea of quantum leaps (an electron could leap from one place to another without visiting the space between) won Bohr the Nobel Prize in 1922. Strangely, the electron, showing a dual nature, sometimes acted like a wave and sometimes like a particle. Heisenberg captured this with the Uncertainty Principle. When observing an electron we can know either the position of an electron or its momentum or pathway but not both. We cannot know or predict where an electron will be but only make a probabilistic assumption. The quantum world even gets stranger with Wolfgang Pauli's Exclusion Principle. Atomic particles can have pairs and when separated they can know what each other are doing. A sister particle will spin to match its twin at the same rate but opposite direction, even if trillions of miles away. Einstein referred to this as "spooking action at a distance," and was bothered that something could outrace the speed of light. Einstein, while contributing a great deal to this field, also had a problem with the notion that quantum world is one of indeterminacy. "God does not play dice," he asserted. Einstein hoped to discover a theory (the Grand Unified Theory) to explain both the world of the very small and the very large. Having two sets of laws in the universe did not make sense to him.

Superstring Theory was also mentioned in this text. At the level of the smallest of the small what was thought of as particles (quarks, leptons) are now understood as vibrating strands or strings of energy that "oscillate in 11 dimensions." Throughout this work Bryson hoped to titillate our imagination and show how science reveals a world of mystery and awesomeness and there is no doubt that he succeeds in this attempt. Does he explain everything? Well, "nearly everything." What a pleasure to read!

Challenging Nature by Lee M. Silver

Lee Silver makes an interesting case that nature is raw, cruel and what the author calls a "nasty mother." An example of the harshness of nature occurred 240 million years ago when almost 95% of all species were wiped out. There is no loving Mother Nature making sure everything works in perfect harmony. It just does not care.

Humanity, on the other hand, does care. So why not pursue techniques, as offered by biotechnology (such as stem cell research), to lessen the blows Nature gives. Eastern cultures tend to fit with this way of thinking more than Western traditions. In the West, there generally is the idea that we are "playing god" when we interfere with Mother Nature. But in the East, where there is no "master plan on the universe," such play with nature is viewed as acceptable. Silver petitions the West to reconsider its stance and to embrace biotechnology and all its benefits. In other words, we should "challenge nature" by utilizing such technology to create a brighter future for all.

Science has so much to offer us. Certainly, we are not at the "end of science" as John Horgan has argued. Instead, science is an ever evolving and enlightening disciple with numerous insights and technologies yet to be had. At the very least it has allowed us to "extricate ourselves from the grip of natural selection."

In the book a section called "Spirits" investigates just how deep and widespread religious beliefs are in the West. There are at least 10,000 different religions worldwide and within Christianity there are about 34,000 Christian denominations. In America 90 plus percent believe in God and about 50% support creationism. Fundamentalism is evidently on the rise. Unfortunately, science is feared by many since their religions offer a contrarian view.

A scientific understanding of the world can be traced back to Aristotle and Democritus with their materialist perspectives. Physicalism, says Silver, is actually the more correct term than materialism since immaterial, massless particles (e.g., photons) needs to be included.

One of my favorite ideas in the book was Silver's explanation how evolution and quantum physics relate. Evolution is driven by random mutations. But how do mutations occur? Most mutations, he explains, are caused by "a high energy cosmic ray (quantum particles) that knocks a single atom of the DNA molecule out of place." This was indeed a brilliant connection between two prominent fields in modern day science. It reminds me of Edward O. Wilson's consilience theory where one field such as physics directly interconnects with another such as biology.

Silver then proceeds to illustrate chaos theory, that a small, seemingly unrelated event can have an enormous effect on the global whole in unimaginable ways. More specifically, an "unpredictable, random quantum event" that results in a mutation can change the course of history. For instance, the development of hemophilia in the blood line of Queen Victoria eventually led to the Bolshevik revolution and the formation of the USSR.

Spooky Physics

This "butterfly effect," as it is sometimes called, peels away at the concept of freewill. Everything is interconnected at the deepest levels (again, consilience theory) and so the idea of one freely choosing one's actions and outcome is very suspect. Couple this with the understanding that at the neuron level there are trillions of neurons in the brain and whether one fires or not makes all the difference in the world on the whole brain. Is there a freely thinking individual controlling that one neuron firing? Certainly not, the evidence indicates. Epiphenomenalism, not Cartesian dualism, is the popular position in philosophy today. The idea that the inner self is an illusion is supported by vast evidence in neurobiology. Francis Crick's *Astonishing Hypothesis* highlights in depth the evidence.

If neurobiology can explain how we think and act can it also explain why is spirituality so ingrained in our consciousness? Yes. The DRD4 gene and the neurotransmitter dopamine (the former processes the latter) can determine one's religiosity. How can the DRD4 gene account for a religious mind? The more active form of the gene the more religious and the least active form the more rational and nonreligious. And increased levels of dopamine also results in a more religious way of thinking. In a study when skeptics were given higher levels of dopamine they leaned toward spirituality. Schizophrenics tend to have 500% more DRD4 than others. Thus, Silver argues that the DRD4 gene must have appeared on the scene as a genetic mutation 30,000 to 50,000 years ago when the religious mind arose in human culture. Perhaps Karl Marx was wrong when he suggested that religion would naturally go away in a just society. Instead, religion could be an innate evolutionary mutation encoded within our DNA but at varying levels. The power of genes keeps religion in play.

One question remains: if the origins of religions can be described as a genetic mutation in the course of human history then why was this mutation naturally selected in the first place? Silver suggests spirituality arose out of an awareness and fear of death. Concepts of life after death relieved anxiety and thus we lived a happier life. As this genetic propensity for spirituality continued, it became amplified with each generation and became the norm. The bottom line: religion was a product of evolution and genetically based. The author calls the genes "spirit genes." And, interestingly, an overdose of them could result in psychosis. Prior to the 1990s scientists viewed religion as a byproduct of culture and not genes. But today the evidence shows otherwise.

Another very interesting section of the book was when the author compares humans with chimps. The 1% difference between us ends up being genes of little significance. We are almost genetic twins with the chimp. Can we produce offspring hybrids together, queries Silver? Most likely, but it would require that the human female carry the fetus and not the chimp (a chimp could not carry such a large offspring full term). Obviously, huge ethical considerations prevent this experiment.

Silver continues to explain that five millions years ago a common ancestor gave rise to humans, chimps and bonobos (pygmy chimps). And as recently as 30,000 years ago we competed with another homo species, Neanderthals, for resources. More remarkable, 18,000 years ago there is evidence of Homo erectus in Indonesia. Homo sapiens probably are responsible for both of their demise.

Why did we develop consciousness as we have it and other creatures like chimps did not? Most likely, Silver says, to "out-compete or kill off cousins who were not equally endowed." How did this mutation occur? As Silver states, "a mutation can be induced by a single cosmic ray that breaks apart a chemical bond between two DNA atoms" and that occurs instantaneously and randomly. The mutation that allowed humans to develop language (and one can argue a form of sophisticated consciousness) is the gene FOXP2. This gene is lacking in the non-human world.

The genome of humans is a subject that receives a lot of attention in this text. Silver clarifies that each human cell contains two sets of about "30,000 genes stored in 46 chromosomes." The genome is all the genetic information within each cell in the human body. While every cell has the same genome, a liver cell, for instance, has the liver portion active within it.

Overall, this was a fascinating book that I highly recommend. Silver's thesis that we need to "challenge nature" (primarily since nature certainly brutally challenges us) was very insightful and appreciated. More than anything, I was especially inspired about the connection between evolution and quantum mechanics and Wilson's consilience theory coming to life.

Schrodinger: Life and Thought *by Walter Moore*

Walter Moore details the life, science and philosophical bent of the famous physicist, Erwin Schrodinger. An only child, Schrodinger was recognized as brilliant even at the age of three. He was always top in his class and eventually became an amazing physicist and mathematician.

In terms of religion, Schrodinger fits in the atheist camp. He even lost a marriage proposal to his love, Felicie Krauss, not only due to his social status but his lack of religious affiliation. He was known as a freethinker who did not believe in god. But interestingly Schrodinger had a deep connection to Hinduism, Buddhism, and Eastern philosophy in general. Erwin studied numerous books on Eastern thought as well as the Hindu scriptures. He was enthralled with Vedanta thought and connected ideas of oneness and unity of mind with his research on quantum physics, specifically wave mechanics.

Schrodinger was almost as much of a philosopher as he was a scientist. While many Western philosophers fascinated him, including Nietzsche, Kant, etc., Schopenhauer was probably the most significant to him. This philosopher shared with Schrodinger an interest in Buddhism and Vedanta thought, which Schopenhauer called atheistic religions. He went on to describe pantheism as "a euphemism for atheism." And Schopenhauer's view of the struggle for existence and the raw, brutal forces of nature seemed to Erwin to accurately depict reality. Spinoza, Einstein's favorite philosopher, was also of great interest to Erwin.

Schrodinger's marriage to Annie was an unusual one. While they remained married throughout their lives and he died with her at his side, he was not attracted to her physically. Both decided to live a more libertine life and engage in discreet affairs. He fathered a couple of daughters with two mistresses. Annie's lover was Hermann Weyl, a scientist and friend of Schrodinger.

What did Schrodinger contribute to physics? Like Einstein he dreamed of discovering a unified field theory but neither scientist were successful there. Instead, Schrodinger made his name in physics and won the Noble Prize for wave mechanics (a wave equation for particles). He was also noted for matrix mechanisms.

Einstein Defiant: Genius vs. Genius in the Quantum Revolution *by Edmund B. Bolles*

While offering a rough outline of Einstein's life, Edmund Bolles focuses on Einstein's resistance to the implications of quantum theory. Einstein did not think that the quantum world was fully understood and that a complete theory was yet to be had. He held faith in the idea that "the universe makes sense and runs on meaningful physical law." The indeterminism of the quantum world did not sit well with Einstein and hence his famous quote, "God does not play dice with the universe." Along the same thought, he expressed, "The Lord is…not malicious." Underlying the indeterminacy of quantum physics, he argued, was an ordered and predictable reality one day to be discovered. The "secret of the Old One," an objectively ordered and comprehensible world, was there to be found. Einstein eventually stood alone in this position; he remained defiant throughout his life. An inner voice, he said, told him that quantum mechanics is "not yet the real thing."

The other genius to counter Einstein was Niels Bohr. Both physicists highly respected and admired each other but could not see eye to eye on this most pivotal research. Bohr, coming from the Copenhagen school of thought, embraced the radical insights of "lawless chaos" and "statistical randomness" quantum theory posed. Causality and meaningful law fell apart at the quantum level, quipped Bohr. The "quantum jump," where a particle leaps from one location to another without following a predicable trajectory or without going through the space in between, was an example of this. Like Bohr, Max Born posited that underneath all the apparent natural laws was only "chaos."

Heisenberg's uncertainty principle accepted by Bohr and Born stated that one cannot know the position and momentum of a particle since they are "exclusive notions." We are left with only probabilistic and statistical interpretations, according to Born. Thus, reality, as classical physics portrayed, now longer fit. Nonetheless, the "correspondence principle" allowed Bohr to use classical ideas to solve modern quantum problems. The bottom line: classical physics did not need to be rejected argued Bohr.

Both geniuses, Einstein and Bohr, also disagreed on the topic of light quanta. Einstein was amazed by the duality of light. His hv refers to the particle-wave duality of light quantum (later knows as photons or even wave packets). Bohr, along with some other physicists, resisted the hv theory but to some degree later came around when the evidence warranted it. "Wavy little chunks of hv" were eventually deemed to be real as the Compton Effect showed.

There is one more area that highlighted the difference between these two thinkers. Einstein loved his philosophers. Just like Erwin Schrodinger, one of his favorites was Schopenhauer. He would study them for entertainment. Bohr, on the other hand, referred to philosophy as "pure drivel."

This book served an excellent read demonstrating how Bohr's "poetic attitude" and Einstein's "realistic" one set the stage for one of the most fascinating debates in the history of physics. And this debate still continues today capturing the attention of scientists around the world.

www.ingramcontent.com/pod-product-compliance
Lightning Source LLC
Chambersburg PA
CBHW080850170526
45158CB00009B/2691